What can you see at Wakehurst?

Too much to see? Do come back – Wakehurst changes with the seasons.

You can find the answers to our puzzles and quizzes on page 46.

House and history

Welcome to Wakehurst Place. Begin your adventure by going back in time.

ACTIVITY

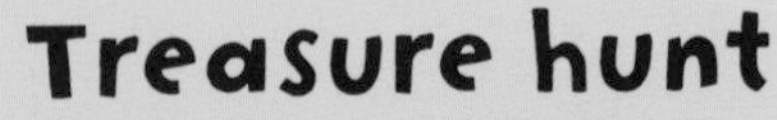
Treasure hunt

Can you find these inside or outside the buildings?

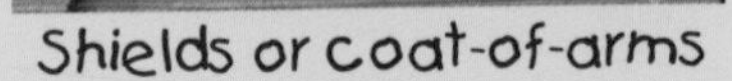
Shields or coat-of-arms

Scallop shell

Angel

Mermaids

Date of the house

Ancient rock

The house is an Elizabethan mansion completed in 1590 for the Culpeper family. It is made of sandstone dug at Wakehurst Place. The sandstone dates back to the time of the dinosaurs.

Plant connections

Nicholas Culpeper (1616-1654) was a relative of the Culpeper family who lived in the house. He wrote books about plant medicines and remedies. One remedy called for small doses of poisonous daffodil bulbs to make an ill-person vomit.

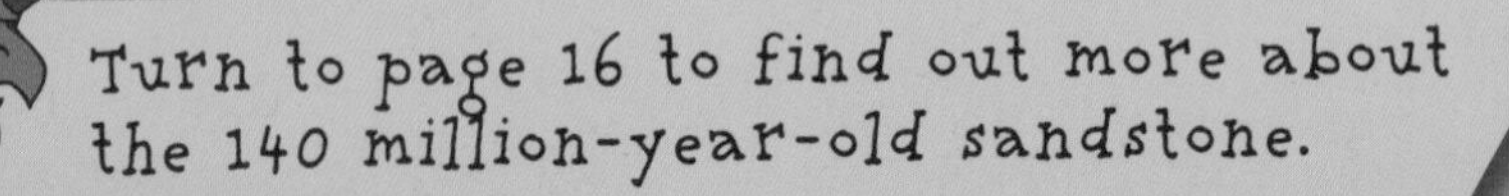
Turn to page 16 to find out more about the 140 million-year-old sandstone.

Great gardeners

The people who lived at Wakehurst Place shaped the gardens you see now with the help of their gardeners. This photograph of the Wakehurst Place gardeners was taken in 1900.

Lords and ladies

Sir Gerald Loder (1861-1936) who later became Lord Wakehurst bought the house and gardens in 1903. Within five years about 3,000 different kinds of plants from all over the world were being grown at Wakehurst Place.

Thank you Sir Henry

Sir Henry Price became the owner of Wakehurst Place in 1935. He and his wife Lady Eve Price enjoyed breeding plants. He left the house to the National Trust, who leased it to the Royal Botanic Gardens, Kew until 2064.

You can find a Pieris named after Sir Henry Price in the Asian Heath Garden.

Hidden history

During World War II a top secret underground room was built at Wakehurst Place. The room was next to an old quarry in the Pinetum. The soldiers had radios to coordinate the resistance if England had been invaded by Germany.

Caught on the hop?
In the late 1800s wallabies were kept in a pen where the toilets are now in the gardens.

Glorious gardens

As you wander outside the Mansion discover...

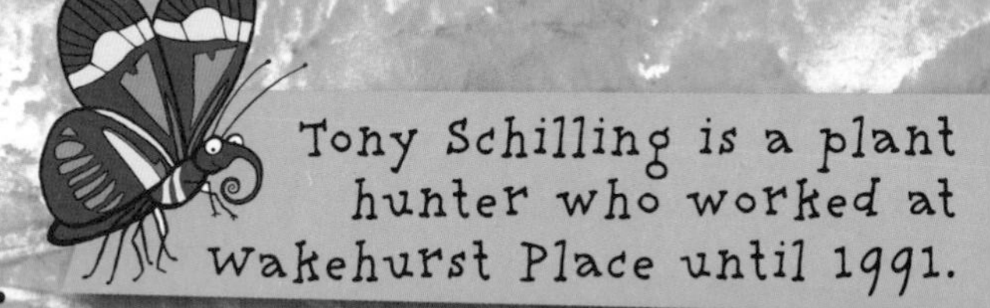

Plants from around the world

You can find plants in the Tony Schilling Asian Heath Garden from the high mountains of Asia. Look for plants from Korea, Japan, Taiwan and China. You can find plants from the parts of the world that lies south of the equator in the Southern Hemisphere Garden.

Water-loving plants in The Slips and beyond

Whiffy plant

The skunk cabbage comes from northwest America. The flowers have an unpleasant skunk-like smell to attract flies. The pollen is carried from flower to flower by the flies. Bears eat the roots to help them do poos after the winter.

Snorkel roots

All roots need to breathe. The swamp cypress grows in boggy places, where the soil has little oxygen. It sends up snorkel-like roots to breathe.

Features in the Bog Garden

Bogs are wet places with soggy soil. The plants around the edge are good places for the young of dragonflies and damselflies to climb up out of the water when ready to turn into adults. At one end is a log stack where frogs and insects can spend the winter.

Blooming bonanza

Our Japanese irises bloom in summer. Can you guess how many different kinds we grow?

Millennium Seed Bank

See scientists at work saving seeds.

What is a seed?

Most plants reproduce by making seeds. Some plants make seeds as a way to survive harsh conditions, such as drought. Seeds also help plants spread to new places. For example, dandelions have fluffy seeds, which are carried away by the wind.

Saving seeds

The Seed Bank stores seeds collected from wild plants from all over the world. Our seed hunters collect seeds from plants at risk, plants only living in one area, and plants useful to people.

Drying and cleaning

We start drying the seeds as soon as they arrive. Debris is removed using a blower or by rubbing them gently against a sieve. The seeds spend up to 30 days in the next drying room. Drying the seeds means they will last longer in the Seed Bank.

Look through the glass windows on the left side with your back to the door. Can you spot the zigzag seed sorter?

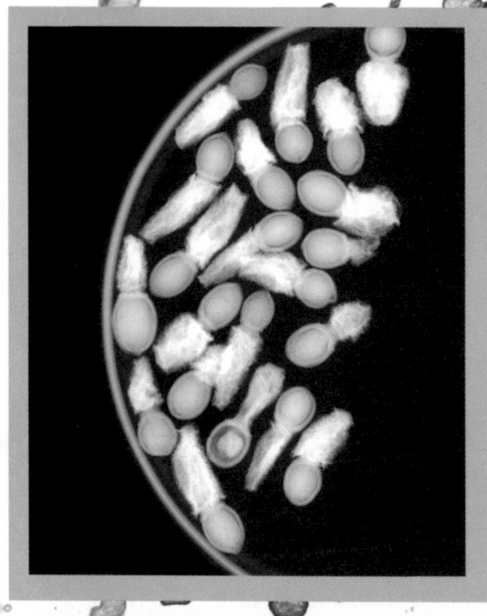

Health check

A small number of the seeds in each collection are x-rayed to see if they are healthy. We check to see if the seeds are fully developed or if any have insects lurking inside them.

In the freezer

The dried seeds are stored in air-tight containers at minus 20°C in the vault under the floor. This temperature is a couple of degrees colder than the ice cream in your freezer. The dried, frozen seeds could now stay alive for hundreds of years.

The vault in the Seed Bank is so big you could fit 30 double-decker buses inside it.

By 2020 we hope to have a quarter of the world's different kinds of wild plants saved in the Seed Bank.

Difficult seeds

Some seeds are hard to keep alive by drying and freezing in the Seed Bank. Seeds from some rainforest plants do not survive if dried. Davidson's plum comes from tropical forests in Queensland, Australia. Its seeds have to be stored in liquid nitrogen at minus 196°C.

Get growing

Some seeds from each collection in the freezer are tested to see if they can still grow. Most seeds need warming up, water and a place to grow. Other seeds have peculiar needs such as a whiff of smoke before they will grow. We put seeds from Tanzania in a flash fire to get them to grow.

Seed hunters

Our scientists search for wild seeds around the world.

In 2007 we banked our billionth (1,000,000,000) seed. It was collected in Mali, Africa from a kind of bamboo.

Safety in numbers

Seed hunters check how many individual plants at any one site are producing seeds. They need to collect plenty of seeds so they can be kept safely for many years. Enough seeds are always left behind so the plant can keep going at that site.

African adventures

In West Africa seed hunters can face dangerous animals such as angry bees or even a charging elephant. Local experts advise making a lot of noise and not putting your hand where you cannot see it to avoid being bitten by a snake.

Identity parade

When collecting seeds, part of the plant is pressed and taken home to check we know what it is. Seed hunters also make notes about where the plant was found.

China find

Scientists thought this Chinese flower had died out 100 years ago. In 2006, over 300 plants were discovered in Yunnan, China. The rare flower grows near to local villages so scientists have collected seeds in case it disappears.

At home

Most of the 1,400 different kinds of British wild plants are now in the Seed Bank including the rare starved wood sedge (below). Seeds dispersed in water can be difficult to collect (above).

Working together

We help over 50 countries to save their seeds. Half the seeds are stored in the country where they were collected and half in the Millennium Seed Bank. We also help to identify plants, set up seed banks and train scientists.

Seeds to the rescue

Seeds can help the world in many ways.

Ready to restore

Seeds can be collected from a place that may become damaged because of an activity, such as mining. New plants can be grown from the seeds to restore the area when the mining is finished.

Robinson Crusoe cabbage

A rare cabbage tree only grows on an island off the coast of Chile. A man marooned there in the 1700s was the inspiration for the story of Robinson Crusoe. Cabbage trees grown from seeds stored in the Seed Bank flowered and produced more seeds in 2004.

Star reborn

The starfruit only grows in a few places in southern England. This rare plant likes the edge of ponds where cattle have churned up the mud. In 2006 some of our seeds and seedlings were put around some ponds to build up the numbers of starfruit.

Better grazing

Land cleared for grazing in Australia has become more salty. Many plants do not like salt. The Seed Bank has given over 150 seed collections to Australian scientists. They are checking whether the seeds could grow in the salty soil.

Seeds of the ribwort plantain

Tree for life

The seed pods of the African locust bean are collected because the seeds are rich in protein and can be eaten by people and livestock. The tree also provides shade, a windbreak and timber for people living in West Africa.

ACTIVITY

Word puzzle

What word means a living thing has died out forever? Fill in the puzzle to find the word in the shaded squares.

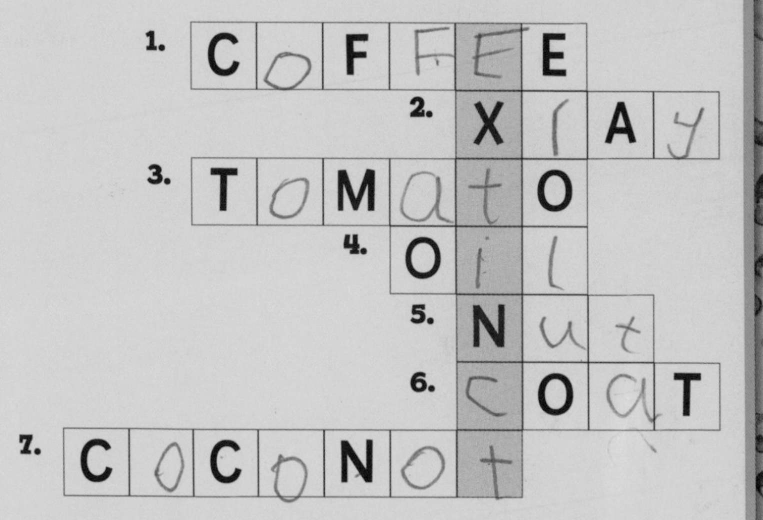

1. What seeds wake you up in the morning?

2. What to do we use to see inside seeds?

3. What seeds can grow after they go through your gut?

4. What liquid comes out when sunflower seeds are pressed?

5. What kind of seed is hard to crack?

6. What covers the outside of a seed?

7. What kind of seed can float across the sea?

On the edge!

See mini UK landscapes outside the Millennium Seed Bank with plants under threat...

...from here to the coast

Ragged robin

Meadow

Hay meadows full of wildflowers are good for insects, such as bees and butterflies. Sadly, meadows have almost disappeared because of the use of fertilisers, weed-killers and stronger grasses.

Visit the Loder Valley Nature Reserve in summer to see our hay meadow in flower.

Our mini downland is too small for sheep to graze instead we cut the grass with scissors.

Cowslip

Chalk downland

Chalk-loving grasses thrive where the downland is grazed by sheep or even rabbits. Grazing keeps back shrubs, which might otherwise outgrow the grasses. Wildflowers, such as cowslips, grow amongst the grasses.

Hoary stock

Cliff face

On the South Coast, the chalk is cut back by the sea into cliffs. Low-growing plants can survive even though it is dry with very little soil.

Sea holly

Shingle beach

Hardy, salt-tolerant plants can grow in the sun-baked shingle. Sea kale and sea holly have waxy leaves to reduce water loss and sun damage. The coast is at risk if sea level rises because our gas-guzzling may warm up the world.

Look at the other landscapes to help you find the answers to the wild word scramble.

Seeds from all the plants you see in these landscapes are kept safe in the Seed Bank.

ACTIVITY

Wild word scramble

Unscramble the words to discover more threatened places.

Many nefs and sharems have been drained for farmland.

Plants growing on lishl and nonstaium are in danger as the world warms up – plants cannot go higher to find a cool place to grow.

Building new houses and planting forests for timber has destroyed nashatheld.

Silcofernd are no longer full of wildflowers because they are sprayed with chemicals.

Making hay

Many people are helping to protect and restore the places where wild plants live around the UK. In the Loder Valley Nature Reserve to get the best growth of wildflowers we make hay by cutting once a year, after they have released their seeds. Sheep are brought in to graze, which makes bare patches where the seeds can grow.

Look inside the Seed Bank to see dryland plants under threat.

Walk in the woods

Head past the Millennium Seed Bank to ....

Put a sticker in the circle next to each tree when you have found it in the woods.

...find beautiful birches in Bethlehem Wood.

Birches come from northern lands. These trees have papery bark that peels off. The oldest known chewing gum was made from birch resin. This sticky fluid oozes out when a tree is cut or damaged. People living in Sweden 9,000 years ago chewed birch resin sweetened with honey.

Can you find a birch with white bark growing in Bethlehem Wood? Fill in the missing letters when you discover its name.

P _ _ A _ _ E _ _

ACTIVITY

Great Stuff

Circle the things birch can be made into.

Track down trees in Coates' Wood.

Alfred Coates was head gardener at Wakehurst Place in the early 1900s.

Sniffles

The oil from eucalyptus leaves is used in remedies to unblock your nose when you have a bad cold. Scrunch up some dead leaves on the ground to smell the oil. Eucalyptus trees come mainly from Australia.

Southern beeches

These trees come from South America, Australia, New Zealand and neighbouring islands. Some kinds are grown for timber.

Ancient survivor

The Wollemi pine was only known from fossils until 1994, when trees were discovered growing in a gorge near Sydney, Australia. There are only about 100 trees in the wild. Growing Wollemi pines in gardens makes sure they will always be around in the future.

Puzzler

The monkey puzzle got its name when first grown in England because its owner joked it would be hard for a monkey to climb up it. In 2009 our scientists collected seeds from wild monkey puzzle trees in Chile to grow in the gardens.

Rock Walk

Roam around the rocks and let your imagination run wild.

Please don't climb or write on the rocks. Our rocks and plants that grow on them are more fragile than they look.

On the rocks

The rocks are sandstone. They are made up of grains of sand left by rivers that flowed into an ancient sea. You can find outcrops of sandstone elsewhere in Sussex and Kent.

Rock shapes

You can see this rock along the path through Westwood Valley. What animal do you think it looks like?

Cracking up

The outer surface of the rocks has a hard coat. If this coat is damaged, the softer inside is easily worn away. The large cracks may have formed when England was covered by ice in the Ice Age.

The rocks you see along this walk were laid down 140 million years ago when dinosaurs roamed the land.

Partners

Lichens grow on rocks as well as on tree trunks. Most lichens like to grow in places where the air is clean and not polluted. A lichen is a partnership between a fungus and a simple green plant called an alga.

Liverwort

Mosses

Low life

Simple plants, called mosses and liverworts, grow on the sandstone rock because it holds water like a sponge.
They like damp shady places.

You can find more mosses and lichens in the Francis Rose Reserve.

Once upon a time

Do you think the tangled roots and strange-looking rocks look like a scene from a storybook? As you go along the path, think of a story. Will your story be about dinosaurs, elves, or wizards...?

Draw a character from your story.

American giants

Take the gentle slope downhill to discover North American giants in Horsebridge Wood.

The General Grant redwood in California is thought to be 1,650 years old. In 1867 it was named after the eighteenth president of the USA.

Terrifically tall

The coast redwood is the world's tallest living tree reaching a height of over 115 metres. Coast redwoods come from the coast of California, USA. Only two of the trees growing here are mature as the others were lost in the Great Storm in 1987.

Big baby

The General Grant giant redwood from California's Sierra Mountains is the world's second biggest tree (by volume) as measured in 2009. Can you find our youngsters grown from one of its seeds? Redwoods take hundreds of years to reach record size. Our redwoods are about 40 years old.

Flaming forests

Redwoods may survive a forest fire if it is not too severe. The thick, spongy bark is fire-resistant. They drop their lower branches so it is harder for flames to leap into the canopy. The cones open in the heat, and the seeds grow well in ash laden soil.

You can also find some young coast redwood trees in the Pinetum.

Cone home

The Douglas fir is one of the world's tallest trees to have ever lived. Native Americans told a story about mice escaping forest fires by hiding inside the fir cones. Look at the ripe cones to see what inspired this story.

Douglas fir cone

Can you spot boxes built for bats? Find out more about our bats on page 23.

Smoking hot

In northwest America, salmon is smoked by burning red alder wood. The wood can also be used to make electric guitars.

Wild food

In the eastern USA, the nuts of the pignut hickory are gobbled up by wild turkeys, raccoons, chipmunks and black bears.

Put a sticker in the circle next to each tree when you have found it in the woods.

Wonder at wetlands

Follow the flow to Westwood Lake.

Discover more water-loving plants and insects in the Bog Garden.

The branches of this willow tree have been cut back or pollarded. The straight shoots can be cut and used as poles.

Water ways

The main stream flows through Horsebridge Wood. The flow is blocked by dams to make Westwood Lake. Water from the Lake is pumped up to the fountain in the Walled Garden. Outlets along the way up from the Lake are used to water our plants.

Walk on water

Go along the boardwalk to cross over the swamp. The plants here like to grow with their roots in water.

Hide and seek

A hide is a place where you can watch birds. The window strips break up your outline, making it hard for the birds to see you. Keep as quiet as you can because they can still hear you. If you are lucky you may see a kingfisher.

ACTIVITY

Be a birder

Draw your favourite bird in the binoculars.

Good growth

Once a year some of our willows are cut back to near the stump. New shoots grow straight up from the cut base. The bendy, young shoots can be made into willow baskets and frames for sweet-peas.

Wild Wakehurst

Keep your eyes open ... you may see some of our wild animals.

If you are seven or older, ask an adult to book a Bat Walk or Badger Watch to see our creatures of the night.

Flutter by

Spring and summer are the best times to see butterflies. You may see colourful butterflies, such as the red admiral, sipping nectar or basking in the sun.

Dragons and damsels

Dragonflies have their wings open at rest. The more delicate damselflies usually close their wings. You can see dragonflies and damselflies flying over water, especially in the Bog Garden and Wetland Conservation Area during the summer.

Blue tailed damselfly

Badgers on a binge

Badgers live in the Loder Valley Nature Reserve. They spend the day in underground tunnels and chambers, called a sett. They come out at night to look for food. They can eat up to 200 worms in a night.

Birds of all seasons

Pheasants were introduced into the UK from Asia perhaps over 1,000 years ago. Birds now breed in the wild as well as being bred and released for shooting.

ACTIVITY

Seeing with sound

Bats are rarely seen in the day like this long-eared bat. These bats use their big ears to listen out for the sounds made by their insect prey.

Dozy dormice

Dormice are secretive animals so you are unlikely to see one. In autumn, they like to eat hazelnuts and blackberries to fatten up. In winter, they go into a deep snooze, called hibernation.

Head for heights

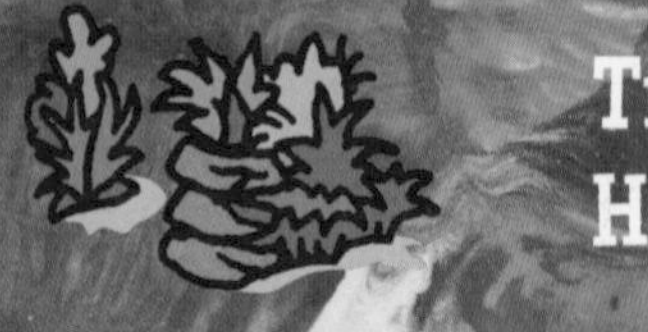

Tramp through the Himalayan Glade.

Keep out

Berberis have sharp thorns and prickly leaves. The prickles may put off grazing yaks in the wilds of the Himalayas. Some kinds of berberis come from Europe. Gardeners sometimes grow berberis to keep burglars away.

ACTIVITY

Complete the picture

Look down when you reach the viewpoint above the Himalayan Glade. Then draw in the missing parts of this scene.

Discover Asian plants in Westwood Valley.

Chinese tulip tree

The tulip tree gets its name because the flowers are shaped like tulips. Our trees are still young so have yet to flower. The trees are under threat in the wild as the forests in China are being cut down.

Bank notes

This pretty, sweet-scented daphne comes from the eastern Himalayas. The bark is used to make high quality paper. It was once used to make bank notes in Nepal.

Mountain blooms

Many rhododendrons come from the foothills of the Himalayas. They grow well in the UK but can be a problem if they take over wild places. Their roots can ruin the sandstone rocks in this part of the country.

First flowers

Magnolias are ancient flowers that date back to a time before bees existed. The flowers do not make nectar. Their pollen is eaten by beetles. They carry some of the pollen from one flower to the next so the plants can set seed.

Pining for pines

Discover pine trees and their cones in the Pinetum.

Most conifers shed and grow some of their leaves all the time – just like your eyelashes.

What is a pine?

A pine is a type of conifer. All conifers have cones. The male cones shed pollen. It is carried by the wind to the female cones. The Pinetum is a collection of pines and other conifers, such as redwoods, firs, spruces, cypresses and cedars.

Needles of stone pine

Flat needles of cypress

Nice needles

Conifers usually stay green all through the year. They have leaves shaped like needles or scales.

Put a sticker next to each tree when you have found it in the Pinetum.

Christmas tree

The Nordmann fir comes from northeast Turkey and the Caucasus Mountains. It is a popular Christmas tree because its needles do not shed as quickly after it has been cut as they do in some other kinds of pine tree.

Coffin tree

The coffin tree comes from Taiwan in Asia. The strong wood is made into coffins for grand funerals. The climate at Wakehurst suits it better than at Kew Gardens where it has to grow inside the Temperate House.

Great Storm

In 1987 a huge storm knocked down more than 80% of the trees in the Pinetum. Wakehurst Place lost over 20,000 trees in total. Since then we have planted many new trees to replace those lost.

ACTIVITY

Make a pine cone animal

You can find pines cones lying on the ground around the coniferous trees in Wakehurst Place as well as in parks and other gardens. Please collect only what you need.

Olly Owl

Do collect fallen petals but please do not pick the flowers at Wakehurst Place. You can find common flowers, such as buttercups and daisies, to pick in rough patches of grass outside the gardens.

Wakehurst at work

Meet the team who keep the gardens and nature reserve going.

Wakehurst Place does not use peat because digging for peat destroys wetlands.

Big job

Our gardeners are kept busy with plenty of weeding, planting, pruning and strimming. In late summer they use motorised platforms to move around outside the Mansion to prune the climbers.

Compost Corner

We make our own compost. Grass cuttings and garden waste make up half the compost. The other half is stinky stable manure. The rows of compost are turned once a week. The temperature inside each row reaches 60°C, which is hot enough to cook an egg. We also recycle waste materials including envelopes and sandwich wrappers.

Beastly bracken

Bracken can choke wild heathers. We help the heather grow in Ashdown Forest by scraping off bracken using a digger. The waste bracken is used as mulch for our rhododendrons and azaleas to make the soil more acid.

Tree people

Climbing trees is all part of the job for members of our tree gang. Dead branches may need lopping off. We have over 8,000 trees to look after in places open to visitors. The gang also mend any holes in the boundary fence to keep out deer.

In the nursery

Some plants are grown to test seeds in the Millennium Seed Bank. Other plants are grown to be part of the living collections you see in the gardens. New plants may be grown from seeds, cuttings or by dividing up underground food stores such as tubers.

Tree protection

When we plant new trees, they have to be protected. Otherwise squirrels and rabbits will tear off the bark to get at the sugar-rich sap. This can kill the young tree.

Baah, baah

Our Southdown sheep are important members of the team. Grazing helps to create flower-rich pastures and meadows in the Loder Valley Nature Reserve. We keep a flock of about 50 sheep. The surplus lambs are eaten in the restaurant.

Working woodlands

Go down to the woods to see how they work.

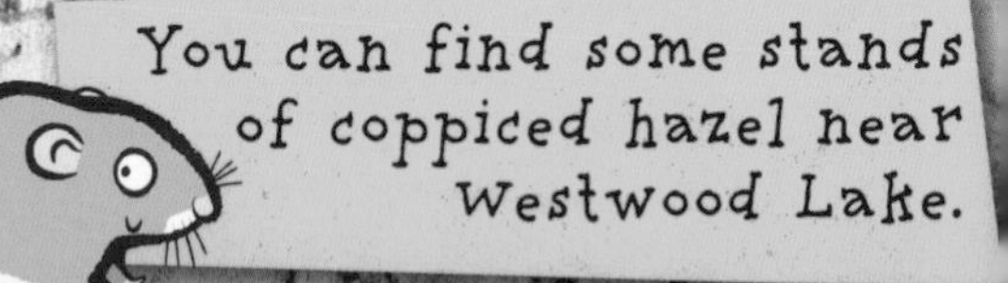

Cutting back

These hazel trees were cut back to the stump over ten years ago. The shoots grow from the stump. This method of cutting back trees is called coppicing. The cut stems are used as stakes, woven fences and furniture. The older and thicker stems can be made into charcoal.

Good for nature

Coppicing lets in light and makes open space for woodland plants and animals to thrive. Woodland butterflies, such as the speckled wood, like the sunny spaces. When the trees mature the hazelnuts are food for squirrels and dormice.

Birch

Ash

Oak

Trees for charcoal

We have coppices of oak, ash and birch trees in the Loder Valley Nature Reserve. Different blocks of trees are cut every 25 years to provide wood so we can make charcoal.

Making charcoal

Stacked up

Some charcoal from the last burn is put into the kiln. This helps to get the new burn started. The charcoal-maker fills the kiln with chunks of wood. The heavy lid is put on top.

Only smoking

The lid is kept open at first to get the fire going then it is sealed with earth and leftover ash. The chimneys in the ground let just enough air into the kiln. If too much air gets in all the wood may turn to ash instead of charcoal.

Sizzle, sizzle

When the smoke turns bluish, the charcoal is ready so the air supply is cut off. The charcoal is allowed to cool. The charcoal is sieved in a rotating drum. The big pieces left inside the drum are bagged and sold.

ACTIVITY

Save the forest

Using our coppiced wood to make charcoal is good for the environment because it...

					2		
					R		
1		A				N	
3		L			I		
					A		

1 across. Saves on ______ dioxide from transporting charcoal from faraway places.

2 down. Means fewer trees are cut down in the ________ forests.

3 across. Makes a place for British ________ to live.

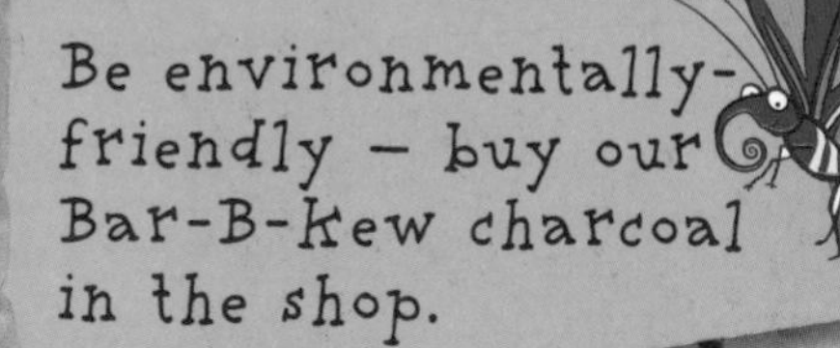

BAR-B KEW charcoal

Spring at Wakehurst

Delight in the first flush of flowers.

Put a sticker in the circle next to each plant when you have found it.

Primroses

March

Primroses brighten up the woods and hedgerows. They produce sticky seeds that ants like to eat. Ants may help spread primroses when they drop some of the seeds.

Daffodils

February to March

Wild daffodil flowers are yellow with a single whorl of petals. We have many different varieties of daffodils bred for both their colour and scent.

What am I?

Early bees like me.

I am about the size of an adult's thumb.

I am usually purple, yellow or white.

Clue — look for me in the gardens on the way to The Slips.

Snakes' head fritillary

April to May

Fritillaries used to be a common wildflower before wet meadows were drained for farmland. The shape of the flower looks like a snake ready to strike.

Waratah

April to May

The name waratah was given to these plants by the Aboriginal people who lived in Sydney, Australia. One of the Australian rugby teams is called The Waratahs. Our waratah comes from the mountains in Tasmania.

Bluebells

April to May

The flower stem of the British bluebell droops to the side its flowers are on. Spanish bluebells from gardens may out-breed the native British bluebell. We dig up and burn any Spanish or cross-bred bluebells so all our bluebells are natives.

In the Elizabethan times, when the Mansion was built, people used the starch from bluebell bulbs to stiffen their pleated collars.

Summer at Wakehurst

Get ready for a blast of warm weather and summer colour.

Himalayan blue poppy

May to June

The plant hunter Frank Kingdon-Ward (1885-1958) sent seeds of the Himalayan blue poppy back to England in 1924. He had many scary adventures during his life of plant hunting including falling off a precipice and experiencing one of the world's strongest earthquakes.

Giant Himalayan lily

June

This lily grows taller than a person. It shoots up from a bulb under the ground. The lily was discovered in the Himalayas in 1821.

St John's wort

July to August

St John's wort is also known as *Hypericum*. Many different kinds are grown here. All have yellow flowers. St John's wort is used in plant remedies to help people feel happier and better able to cope.

Put a sticker in the circle next to each plant when you have found it.

What am I?

I am a tree.

My leaves are green.

My name describes my white 'flowers' that appear for a few weeks in April or May.

Clue — look for me in The Slips

Wild orchids

June to July

Some people think of orchids as potted plants with showy flowers or tropical glasshouse blooms. Some beautiful orchids, such as the common spotted orchid, grow wild in the UK.

Stephen

Tara

Ginger-lily

July to September

Imagine having a plant named after you. The plant collector Tony Schilling found two varieties of ginger-lily in Nepal. He named them after his children. 'Stephen' was collected in 1966 and 'Tara' in 1972.

Autumn at Wakehurst

The days are getting shorter and leaves are changing colour.

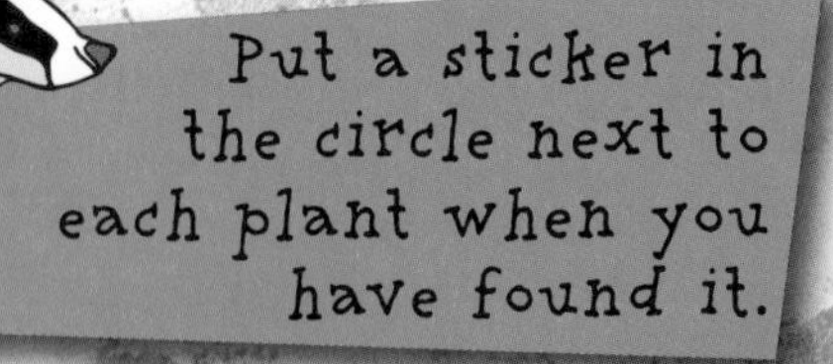

Put a sticker in the circle next to each plant when you have found it.

Japanese maple

September to October

The Japanese maple has been grown in gardens in Japan for centuries. There are thousands of different varieties. The leaves of this one turn brilliant orangey-red.

Cherry birch

September to October

The leaves of this birch turn bright yellow. The name comes from the reddish coloured bark of younger trees. The twigs contain wintergreen. Small amounts of this toxic oil can help to relieve tired muscles.

Mountain ash

September to October

There are several hundred different kinds of mountain ash. These are also called rowan trees. Their scientific name is *Sorbus*. This kind of mountain ash comes from the Himalayas. The bright red berries are eaten by birds.

Katsura

August to October

The leaves of this Japanese tree turn a reddish colour in autumn. Find some dead leaves beneath the tree. If you crush the dead leaves, they smell of candy floss.

Kew keeps over one million dead fungus specimens. It is the largest collection in the world.

Fly agaric

September to October

You can find this fungus next to birch or pine trees. The underground part of the fungus is a mass of threads, which wrap around the tree's roots. The poisonous fungus gets its name because it was used to kill flies.

What am I?

I am a tall tree.

My leaves are like the shape of your hand.

You can play a game with my shiny brown seeds.

Clue – look for me on the lawn near the Mansion.

Winter at Wakehurst

Warm up with a wintertime walk.

Find out more about giant redwoods on page 18.

Skimmia
December to January

You need two plants to get berries. The male plant has flowers that shed pollen. When pollen is carried by bees to the female plant it can then produce the berries. Skimmia are popular garden plants. There are four different wild kinds that grow in Asia.

Snowdrops
January

Snowdrops grow from bulbs underground. They grow in woods and other places where there is shade in summer to keep the bulbs cool. A substance first discovered in snowdrop bulbs is used to treat Alzheimer's disease that makes older people suffer from memory loss.

Christmas giant
December

Our Christmas tree is a giant redwood about 35 metres tall. The Wakehurst Place tree team use cranes to decorate it with 1,800 low-energy lights. It takes a day to put up the lights, which are wrapped around ropes hung from the top of the tree.

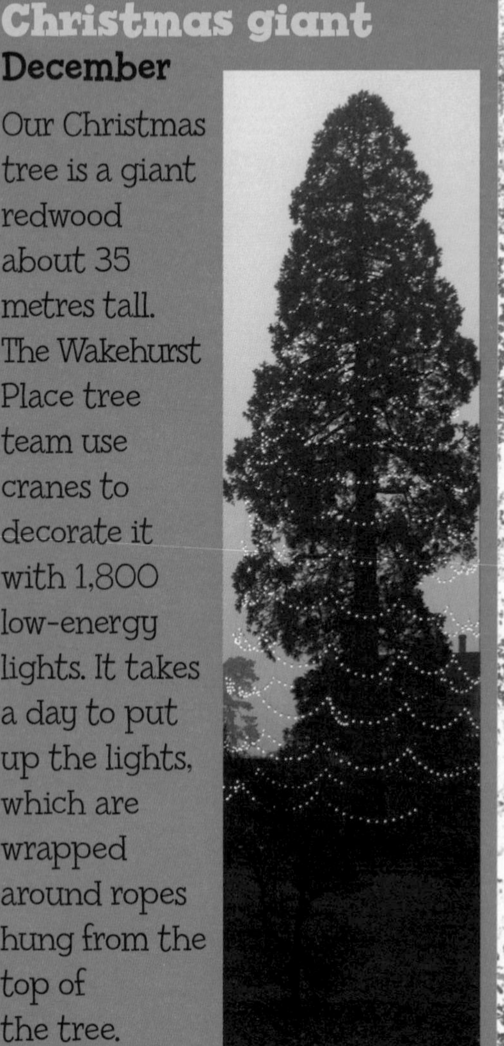

Cyclamen

January to February

Cyclamen come from countries around the Mediterranean Sea. We planted over 30,000 tubers of two different kinds. A tuber is a swollen part of the stem that stores food. One kind flowers in the autumn. The other you see here flowers in winter.

Put a sticker in the circle next to each plant when you have found it.

Dogwood

November to February

The bright stems of the dogwood are easy to see after the bush has lost it leaves. The stems are cut every few years because the older ones lose their colour. This kind of dogwood comes from Siberia.

Witch hazel

December to February

Sniff the flowers for a delightful scent. Extracts from the bark and leaves are used to treat bumps and insect bites.

What am I?

I grow on trees and suck up their sap.

My berries are white.

Keep clear of me, if you don't like to be kissed.

Clue – you can find me on a maple tree near the Mansion Pond.

Seasonal activities

Things to do...

...in spring

Make a bangle

You will need: double-sided sticky tape, scissors, strong paper, fallen petals and leaves

Cut a strip of strong paper about three centimetres wide and long enough to wrap around your wrist.

Stick double-sided sticky tape onto the strip.

Peel off the cover from the sticky tape.

Wrap the strip around your wrist and stick the ends together.

Find fallen petals and leaves to stick on until the bangle is covered.

...in summer

Make a butterfly

You will need: petals and leaves of different sizes, twigs, paints, clear or double-sided sticky tape

Collect fallen flower petals, leaves and twigs.

Arrange them on a piece of paper.

Stick them down and paint a background for your butterfly.

...in autumn

Leaf prints

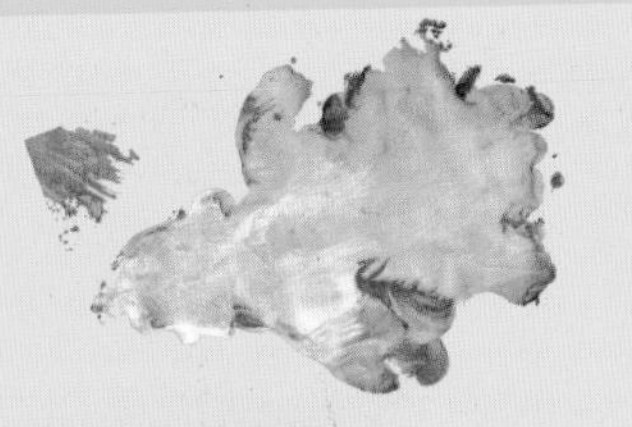

You will need: fallen leaves, paper, paints

Collect fallen leaves of different shapes.

Paint one side of the leaf.

Press your leaf against the paper.

Peel back the leaf to reveal your print.

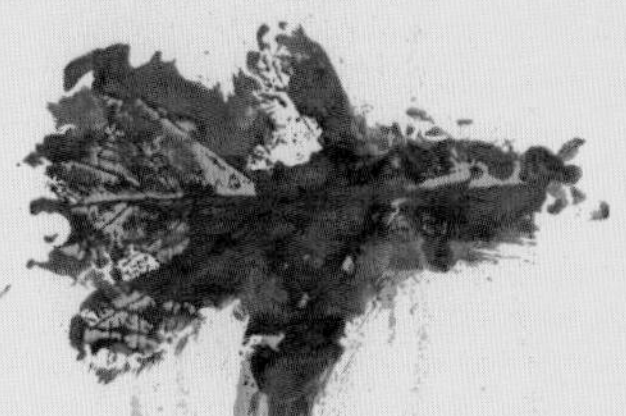

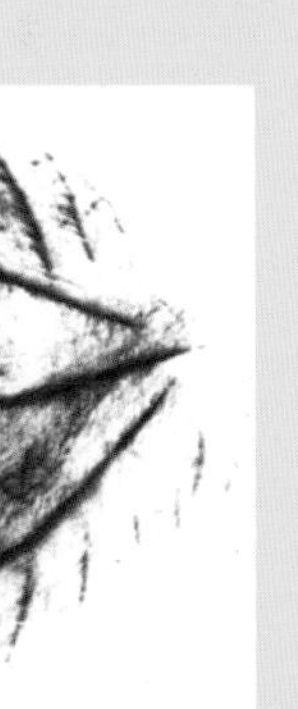

...in winter

Don't forget to write Seasons Greetings!

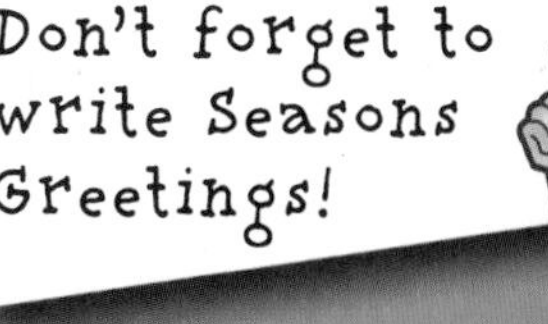

Festive cards

You will need: skeleton holly leaf, glue, glitter, coloured pens, A4 card, old newspaper

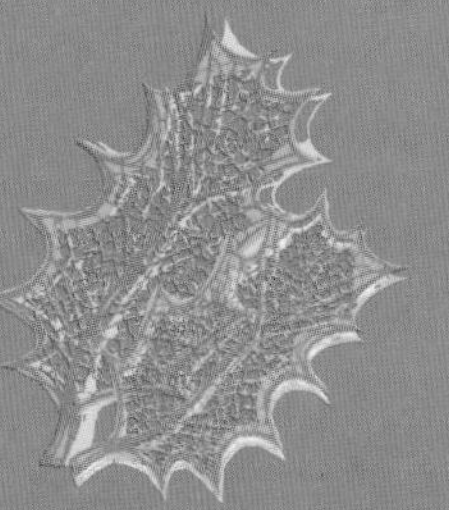

Look on the ground for a holly leaf with a lacy skeleton.

Flatten the leaf between sheets of newspaper.

Put books on top of the newspaper to press

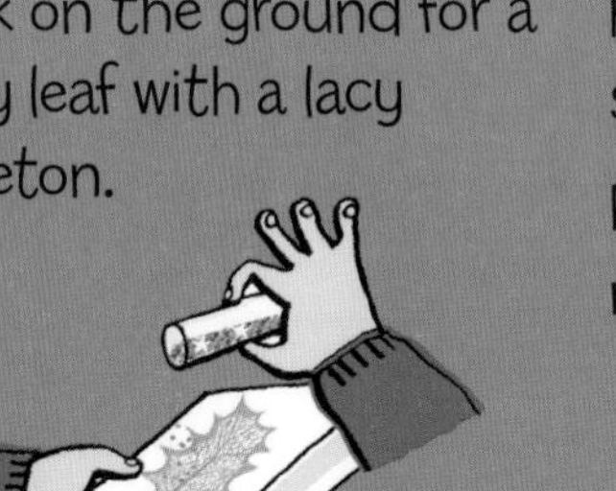

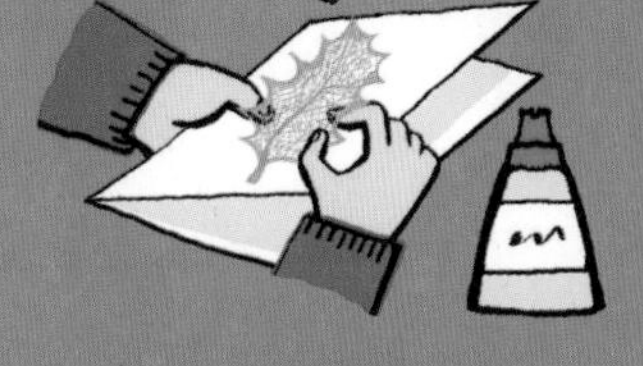

Fold the card in two and stick your holly leaf on the front.

Put some glue on the leaf and sprinkle with glitter.

Tree activities

Things to do with trees.

The Douglas fir in Westwood Valley is the tallest tree in Wakehurst Place. Over 43 metres tall, it is almost as high as the Statue of Liberty.

What is a tree?

A tree has a single woody stem that lasts for many years. It grows at least three metres tall, which is about one and half times the height of a tall man.

English oak

How high is that tree?

You will need: pencil, paper, tape measure

- Walk away from the tree.
- Bend over and look back between your legs at the tree.
- Move closer or further away until you can see the top of the tree where your legs meet.
- Leave a pencil on the ground to mark the spot.
- Count how many paces you make to get back to the base of the tree.
- Write down this number with the pencil collected from the marker spot.
- Measure the length of your pace using a tape measure.
- Multiply the number of paces you did by the length of your pace to get the approximate height of the tree.

My tree was roughly .... metres tall.

Draw your tree here.

Make a bark rubbing

Bark is the outer layer of woody plants including trees. The colour and pattern of the bark can help you identify a tree. To make a record of the pattern take a rubbing of the bark.

You will need: strong paper, wax crayons or chalk

Place your paper on the bark. Rub the crayon or chalk lengthwise across the paper.

Bark rubbing

ACTIVITY

Match the leaf to the tree

monkey puzzle

silver birch

oak

holly

giant redwood

ACTIVITY

Bark colours

Tick the box when you have found a tree with this coloured bark.

 White  Red Green 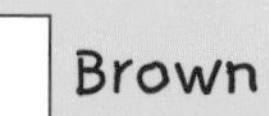 Brown

Here are examples we have found in the gardens.

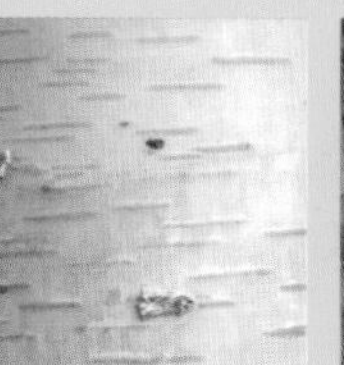
birch

Pacific madrone

moosewood

oak

Seed activities

Things to do with seeds

Seedbank quiz

Can you pick the correct answers?

Clue – you can see the seeds on display in the glass showcase in the Millennium Seed Bank.

1. What is the name of the world's largest seed?
a. Breadfruit **b.** Double coconut **c.** Elephant seed **d.** Giant's droppings

2. What seeds are the main source of food for about half the people in the world?
a. Apple seeds **b.** Groundnut **c.** Wheat **d.** Rice

3. What seeds are used to make your jeans?
a. Cardoon **b.** Flax **c.** Woollybutt **d.** Cotton

4. What are some of the other names for groundnuts?
a. Monkey nuts **b.** Big nuts **c.** Earthnuts **d.** Peanuts

5. What seeds are used by people from the Kalahari Desert, Africa to treat sore joints?
a. Devil's claw **b.** Brazil nut **c.** Grapefruit **d.** Self-heal

Find out about collecting seed from common wild plants on The Great Plant Hunt www.greatplanthunt.org

Be a seed hunter

Try collecting seeds in your garden or window box, instead of buying a new packet for next year. Wallflowers are easy to grow.

You will need: envelope, plastic pot

Look for ripening seed pods that are brown and starting to split.

Break off the stems to collect the seed pods.

Leave the seed pods in a big open envelope to dry in a warm place.

Shake the dried pods to release the seeds.

Put your seeds into a container and store in a cool dark place until next year.

Make a seedy picture

You will need: square of material or paper, different kinds of seeds, glue

Look at our seed picture then have a go at making your own. You can use seeds that you collected on a walk, such as acorns, sweet chestnuts and conkers. These big seeds are hard to stick down so why not take a photograph of your picture to keep.

If you are at home, you can raid the kitchen cupboard for seeds, such as rice, beans, sesame seeds and poppy seeds.

Place the glue on the material or paper.

Sprinkle small seeds onto the glue and gently press down bigger seeds.

You can buy a Mini Seed Bank for your seeds in our shop.

Match the seed to the product

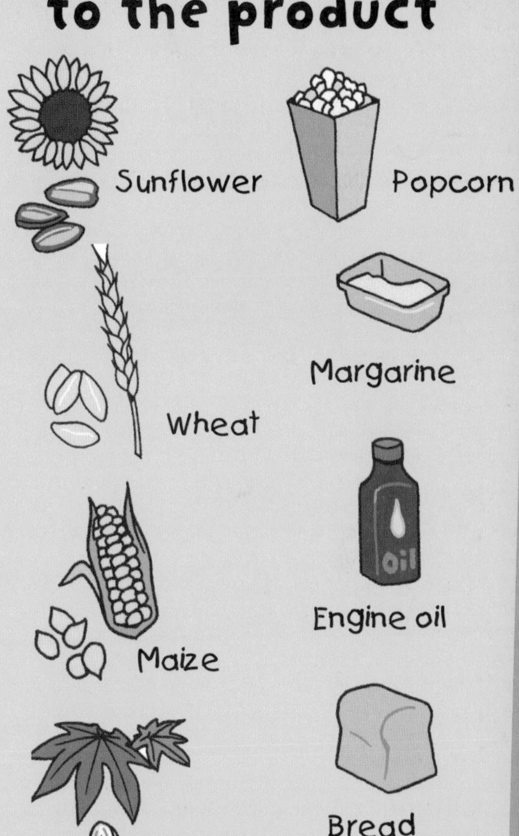

Lotus seeds thought to be over 1,200 years old grew successfully after they were found in the bottom of a Chinese lake.

Answers

Page 2 Treasure hunt
Shields or coat-of-arms - above the largest fireplace.
Scallop shell - on the lead drainpipe outside.
Mermaids - on the ceiling in the Blue Room.
Angel - in the Chapel high on the wall.
Date of the house - 1590, above the door in the Chapel.

Page 4 Where in the world? Match the plant to where it comes from.
China Gingko - female trees have smelly seeds.
Taiwan Spiraea - bees visit flowers in summer.
Japan Pieris - flowers in spring.
Korea Fir - cones are greyish blue before they ripen.
New Zealand Toothed lancewood - the young trees have leaves with sharp edges.
Australia Banksia - named after Sir Joseph Banks (1743-1820) who had plants sent to Kew.
South America Firebush - named for its flaming red flowers.
South Africa Red-hot poker - pollinated by sunbirds in South Africa.

Page 5 Iris Dell blooming bonanza
The Iris Dell has 60 varieties of Japanese iris.

Page 11 Word puzzle
1. What seeds wake you up in the morning? COFFE**E**
2. What to do we use to see inside seeds? **X**-RAY
3. What seeds can grow after they go through your gut? TOMA**T**O
4. What liquid comes out when sunflower seeds are pressed? O**I**L
5. What kind of seed is hard to crack? **N**UT
6. What covers the outside of a seed? **C**OAT
7. What kind of seed can float across the sea? COCONU**T**

Answer: **EXTINCT**

Page 13 Wild word scramble
Many **fens** and **marshes** have been drained for farmland.
Plants growing on **hills** and **mountains** are in danger as the world warms up - plants cannot go higher to find a cool place to grow.
Building new houses and planting forests for timber has destroyed **heathlands.**

Cornfields are no longer full of wildflowers because they are sprayed with chemicals.

Page 14 Great stuff
Chair Broomstick Charcoal Paper
The birch with white bark in Bethlehem Wood is called POLAR BEAR

Page 31 Save the forest
1. Saves on **carbon** dioxide from transporting charcoal from faraway places.
2. Means fewer trees are cut down in the **tropical** forests.
3. Makes a place for British **wildlife** to live.

Page 32 Buzzy Bee activity - I am a crocus.

Page 35 Buzzy Bee activity - I am a handkerchief tree.
My white 'flowers' are in fact a special type of leaf called a bract.

Page 37 Buzzy Bee activity - I am a horse chestnut tree.

Page 39 Buzzy Bee activity - I am a mistletoe plant.

Page 43 Match the leaf to the tree and Bark colours

monkey puzzle silver birch oak giant redwood holly

Page 44 Seedbank quiz
1. b. Double coconut is also known as Seychelles nut palm and coco-de-mer.
2. d. Rice
3. c. Cotton
4. a. Monkey nuts c. Earthnuts d. Peanuts
5. a. Devil's claw

Page 45 Match the seed to the product
Sunflower - Margarine Castor oil - Engine oil
Wheat - Bread Maize - Popcorn

Discover more

Get serious

If you like looking at nature, ask an adult to take you on a walk through the Loder Valley Nature Reserve. There are woodlands, wetlands and meadows to explore. To open the gate to the reserve, they will need to ask for a code from the Visitor Centre.

Making magic

We can help nursery teachers plan a magical day out at Wakehurst Place. Our green fairy can help inspire children about the wonder of plants.

Visit Kew's www.greatplanthunt.org to find out how you and your school can get out and about, and excited by nature. There are loads of fascinating things to do and some fun games too.

Following Darwin's footsteps

Charles Darwin (1809-1882) was interested in nature from a young age. He became one the world's most famous scientists. His big idea was how living things changed through the ages. As a young man, he visited the Galapagos Islands, home of the giant tortoise. You can find out more about Darwin and how to study plants on Kew's www.greatplanthunt.org.

First published in 2011 by
The Royal Botanic Gardens, Kew,
Richmond, Surrey TW9 3AB, UK

Written by Dr Miranda MacQuitty
Illustrated by Guy Allen
Designed by Louise Millar
Development editor: Lydia White

Additional research and activities by Susan Allan, Sarah Bell, Chris Clennett, Astrid Krumins

ISBN 978 1 84246 415 1

www.kew.org

Printed in China by Midas Printing (within a factory that has ISO 14001 accreditation, the internationally recognised standard of environmental responsibility). The paper used in this book contains material sourced from responsibly managed and sustainable commercial forests, certified in accordance with the FSC (Forestry Stewardship Council).

The following people have provided invaluable help and advice: Francis Annette, Stephen Cook, Gina Fullerlove, Andy Jackson, David Marchant, Andy Marsh, Stephanie Mills, Christine Newton, Jo O'Byrne, Iain Parkinson, James Pumfrey, Paul Reader, Steve Robinson, Mary Smith, Roger Smith, David Waldon, John Withall, Rob Yates.

With many thanks to pupils at St Peter's Ardingly for all the fantastic artwork and posing for photos that appear in the book: Francesca Allen, Mathew de Ath, Lami Balogun, Jack Battey, Owen Blunden-Dell, Josh Brown, Joseph Crutchley, Jodie Davidson, Luciana Denney, Emily Fairhall, Olivia Farmborough, Esther Forgie, Harry Hudson, Jack Jackson-Humphrey, Samuel Lambert, Elliot Lewis, Blair McIntosh, Paul McNamara, Rory McNamee, Maddie May, Gianluca Pasquini, Mogi Salamat, Libby Serdink, Lillie Shelley, Daniel Shillito, Trudy Louise Turtle, Hope Warren, Oliver Woolgrove. Teacher Jill Maskell. Adult helpers: Nicky Brown, Hannah Felton, Dionne Flatman, Alex Rumble Volunteers : Larry Halley & Margaret Mann

We would like to thank the following for providing photographs and for permission to reproduce copyright material: Heather Angel 44; Chris Clennett 4, 9, 12, 15, 16, 17, 18, 23, 27, 30, 35, 36; Emma Dodd 41; Peter Gasson 17; Jim Holden inner front flap; Images & Stories/ Alamy 22; Miranda MacQuitty map, 4, 5, 12, 15, 16, 20, 21, 25, 26, 30, 44; Richard Manuel/ www.photolibrary.com 10; Louise Millar 5, 40; Olaf Protze 10; Steve Robinson 22, 23; Shui Yumin 9; all other photographs by Andrew McRobb.

For information or to purchase all Kew titles please visit www.kewbooks.com or email publishing@kew.org

Kew's mission is to inspire and deliver science-based plant conservation worldwide, enhancing the quality of life.

Kew receives half of its running costs from Government through the Department for Environment, Food and Rural Affairs (Defra). All other funding needed to support Kew's vital work comes from members, foundations, donors and commercial activities including book sales.

Put a sticker next to each plant in the book when you have found the plant in the gardens.
Southern beech
Cyclamen
Himalayan blue poppy
Cherry birch
Wollemi pine
Himalayan lily
Pignut hickory
St John's wort
Mountain ash
Waratah
Ginger-lily
Giant redwood
Fly agaric
Bluebell
Dogwood
Birch
Daffodil
Skimmia
Red alder

Monkey puzzle
Japanese maple
Snowdrop
Douglas fir
Eucalyptus
Witch hazel
Nordmann fir
Katsura
Snake's head fritillary
Coast redwood
Primrose
Wild orchids
Coffin tree